Today I walked into a cloud of Dragonflies...

A Book of Inspiration and Quotes
by Suzan David

Today I walked into a cloud of Dragonflies...
A Book of Inspiration and Quotes by Suzan David

ISBN-13: 978-1723317019
ISBN-10: 1723317012

Please visit us online at SuzanDavid.com.
We would be happy to help you get healthy. It's not too late
to have an amazing life!

The quotes by famous people were borrowed according to
fair use law and attributed.
The photo of Suzan David on the back cover was taken and
edited by Walter Johnson Jr. and is used by permission.

Cover painting by Suzan David

Published by Monument Press 2018
Printed by CreateSpace

monument
P R E S S

A division of
Bluff City Communications Inc.
books@bluffcitycomm.com

About 3 months ago I started seeing Dragonflies everywhere. They were always around. They landed near me. Book covers with a Dragonfly would catch my attention. It was kind of weird.

I was not really happy about this. Because even though they are kind of cool, they are – you know – bugs. YUCK. But I paid attention, and soon one of my friends told me they were the symbol of transformation. Transformation?? Wow, I desire transformation! I am ready to step into my greatness. I am ready to fly!

So why are Dragonflies significant? A couple reasons…

- They live most of their lives as a nymph. Immature, getting ready. Immature. (Sound familiar?)

- Then they transform into an adult Dragonfly and have only a few months to soar. So they have to fully enjoy every minute of their transformation. They live in the moment. They do what they want and make decisions quickly.

Yea… speaking as someone who has been a nymph for most of my life, I'm ready for that transformation.

So, first let's talk about the obvious. Let's get our bodies and minds in shape. These lovely bodies are our vehicles. Our wings. Our mind chooses the shape and health of that body.

So here are some notable people talking about your mind-set and your body. Pay attention to this great wisdom!

" Smile in the mirror.

Do that every morning and you'll start to see a big difference in your life. "

~Yoko Ono

"Every day put on wild music that makes you happy and dance like a crazy person.

Do this first thing in the morning and you will have a great day!"

~Suzan David

"You can do ANYTHING you want to do. This is your world.

~ *Bob Ross*

" When you do things
from your soul, a river
moves through you.
Freshness and a deep
joy are the signs. "

-Rumi

"
Change your thoughts
and you change
your world.
"

~*Norman Vincent Peale*

" Only I can change my
life. No one can do
it for me. "

~Carol Burnett

"You never fail
until you stop trying.

-*Albert Einstein*

Stop making excuses. You can do anything!

-Katia Beauchamp

Don't stop when
you're tired;
stop when
you're done.

~Marilyn Monroe

" Either you run
the day, or the
day runs you. "

~Jim Rohn

"I am not a product of my circumstances. I am a product of my decisions.

~Stephen Covey

Challenges are what make life interesting and overcoming them is what makes life meaningful.

~*Joshua J. Marine*

Life isn't about waiting for the storm to pass... It's about learning to dance in the rain.

~*Vivian Greene*

"If you can dream it,
you can do it.

~*Walt Disney*

"Be the best you,
so you can show
someone else
the way.,,

~Paul Glinn

I'm 58. That's nearly 60!! I have wrinkles by my eyes and my hair is naturally gray. But I have lost some weight and I have regained my health and energy. In celebration I do wonderful things to celebrate life! One of those things was paddle-boarding with a friend. You stand up on a kayak-like board and paddle from the standing position. It's wobbly. I did fall onto the board a couple of times (but not into the water!) and after sitting for a while a Dragonfly landed by my foot. I looked at it. It just sat. I looked around. I inhaled the lake air, and looked at the clouds and the blue sky and the sun. I felt the water with my finger-tips and felt the breeze on my face. I was IN THE MOMENT. I was TRANSFORMED. Because I stopped to look at the Dragonfly. I'm so grateful for that moment!

Let's live in the moment!

> Make the most of
> your day! Time is
> precious.

~*Tess Bear*

"The will to win, the desire to succeed, the urge to reach your full potential... these are the keys that will unlock the door to personal excellence.

~*Confucius*

"Remember that you were once a brave little girl who thought the whole world was open to you. It is still true! Find that little girl and reconnect. She knows the way!"

~Suzan David

" The sky is not
the limit.
Your mind is. "

~ *Marilyn Monroe*

"If you want to live a happy life, tie it to a goal, not to people or things."

~Albert Einstein

"Listen to the voice in your head to see what it's telling you. Then tell it to stop lying.

Change the message to serve you!"

~Suzan David

" The best way to
predict your future
is to create it. "

~Abraham Lincoln

Optimism is the
faith that leads to
achievement. Nothing
can be done without
hope and confidence.

~*Helen Keller*

Change your story change your life.

Basically, that's what it is.

~Deepak Chopra

"Keep at it. You never know when what you desire will appear."

~*Sharon VanBrunt*

"Be thankful even for the bad days. Some souls don't even get any more of these. While there is breath, there is hope."

~*Brent David*

To transform your life, you must make intentional choices. You must decide what you want and then go get it. You might need to change your body. You might need to chase a few dreams. Whatever it is you decide to do, you must pursue and never give up. A Dragonfly only has a few months to fly. We may have a long time and we may have a short time. But let's really go for it!

One important thing to remember is that flying can be difficult. You are going to face set backs and you are going to have to push through.

These amazing quotes encourage us to make that happen!

"You can cry again or
You can try again.
Choice is yours."

~Soham Mondal

"Make the choice that you want to do something and then do it. Don't try if you aren't all in, because you won't be successful.

Start when you are truly ready. "

~Lisa Powell Lafarlette

"It's not whether you get knocked down, it's whether you get up."

~Vince Lombardi

" Don't compare your beginning to someone else's middle. "

~Sandra Yancey

Action is the most
important as well
as the most difficult
aspect to any plan.

~Beverly Schaefgen

" Do it, and then you
will feel motivated
to do it. "

~Zig Ziglar

"You see, in life, lots of people know what to do, but few people actually do what they know. Knowing is not enough! You must take action."

~Tony Robbins

"You have it in you.

You don't need to wear it on your sleeve, just be calm and come to an agreement with yourself, this is me, it is who I am and I will think above, around and in ways the others won't. Creativity is as important as honesty."

~Bob Sebastian

"You can start with
nothing. And out
of nothing and out
of no way, a way
will be made."

~Michael Bernard Beckwith

I walked into a cloud of Dragonflies today. As I walked, I wondered if I could walk away from them, but they followed me. Hundreds of them. They did not land on me and they weren't intrusive, they were just there. Everywhere.

I understood in that moment that NOW is the time to change.

Today, they are visiting YOU! Let's transform together!

Never give up. You CAN do this, and you will be so much happier when you do!

"When obstacles arise, you change your direction to reach that goal. You do not change your decision to get there."

~Zig Ziglar

Never, never,
never give up.

~*Sir Winston Churchill*

> Do or do not.
> There is no try.

~Yoda

"Face your fears with the truth, that they are all in your mind, and they will lose their power over you."

~Jen Sincero

"Our greatest weakness lies in giving up. The most certain way to succeed is always to try just one more time."

~Thomas A. Edison

" Success is not final,
failure is not fatal,
it is the courage to
continue that counts. "

~Sir Winston Churchill

Your Healthy Life

- Put your mind in a great place every morning.

- Create routines that support your decision to eat healthy.

- Plan snacks and meals in advance.

- Find a coach or weight loss partner.

- Do self-care. Show your body love. You are regaining your health to honour your body, and it will support you.

- Be kind to yourself.

- Surround yourself with people who support you.

- Move more. Your body is a wonderfully made vehicle for your soul, and it craves movement.

- Take your measurements when you begin. Not all health victories are related to the scale.

- Take "before" and "during" photos.

- Congratulate yourself with every victory – no matter how small.

- Remember that you are amazing!

www.ingramcontent.com/pod-product-compliance
Lightning Source LLC
Chambersburg PA
CBHW071243220526
45468CB00002B/990